LAURELWOOD SCHOOL
SCIENCE LAB
955 TEAL DRIVE
SANTA CLARA, CA 95051

LAURELWOOD SCHOOL
SCIENCE LAB
955 TEAL DRIVE
SANTA CLARA, CA 95051

Designed and produced by
Aladdin Books Ltd
70 Old Compton Street
London W1

First published in
The United States 1982
by Franklin Watts
730 Fifth Avenue New York
New York 10019

U.S. SBN: 0-531-04379-7

Library of Congress
Catalog Card No: 81-71546

Printed in Belgium

A typical street in any Western city center

HOMES & CITIES

COLIN MOORCRAFT

FRANKLIN WATTS
London · Toronto · New York · Sydney

Ray Dafter *is Energy Editor of the Financial Times. In 1978 he spent a year doing research at Harvard University, and lecturing in the United States. He has written three books on Energy, broadcasts on radio and television, and contributes to a number of publications.*

Until recently mankind had taken energy for granted. It was always there – why bother about what it is or where it comes from? But today people are worried. Suddenly we discover that we have been using up important stores of energy – oil, natural gas and coal – far too quickly. We must all of us think very seriously: where will our energy come from in the future? For there is nothing in the world that is not affected by energy – or the lack of it – as this book and others in the series will show.

Our ancestors, living in an underpopulated world, had other concerns. One result was that throughout history, people steadily deserted the countryside in favor of towns. The advantages – improved trade and communications – seemed obvious.

However, as HOMES AND CITIES clearly shows, our present day urban lifestyle has some drawbacks. Cities are noisy, and often congested and above all else our modern homes and cities now require vast amounts of energy in order to function properly. It is important for us all to question how we can continue to enjoy modern comforts but without such heavy demands on our precious fuel resources. This book makes some important suggestions.

Ray Dafter: *Consultant Editor*

This is a condensed view of the world in which we live – and every single thing in it requires energy. Even stationary things, such as buildings, bridges and docks take energy to build, and need further energy inputs to function and be maintained. Transport, industry, our homes, cities, and the very food we eat simply could not exist without it. We obtain energy from very few sources, and unfortunately it is often locked up below ground. Since we have to use energy in order to make more, it is a cycle as complete as any ecosystem – with one exception, nature reabsorbs all her waste products. We don't.

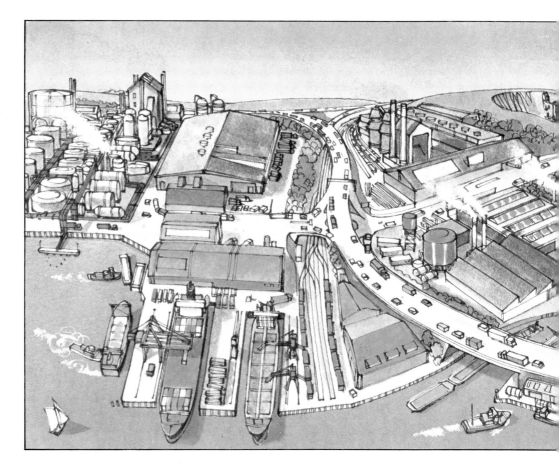

Contents

Art Director	Charles Matheson
Art Editor	Ben White
Editor	James McCarter
Designer	David West
Typographic Design	Malcolm Smythe
Research	Dee Robinson
Illustrators	Denis Bishop
	Peter Hutton
	Industrial Art Studio
	Jim Robins

For the purpose of this book:
A billion is one thousand million. A trillion is one million million.

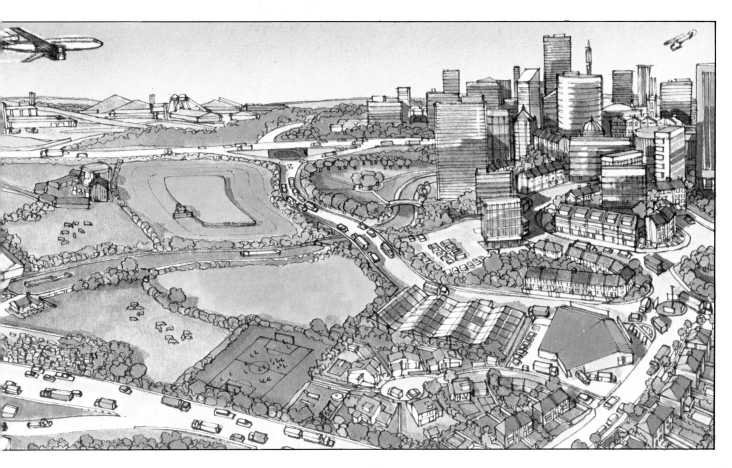

What is a city?

Cities are one of man's most complex developments, providing housing, employment, transport and entertainment to their millions of inhabitants. They support a vast range of different activities in a way that smaller communities would find impossible. City dwellers enjoy a wide field of choices in every aspect of their lives.

An imaginary journey outward from the city center into the surrounding countryside gives an indication of the many types of buildings, either homes or places of work, that go to make up the typical city of an industrialized country.

The city center itself is dominated by commerce, administration and entertainment. Banks and insurance companies have their headquarters here in towering skyscrapers. City hall and federal government buildings will be found close by. Shops, restaurants, bars and theaters line the sidewalks.

In the area immediately circling the city center some of the city's oldest housing is usually found. This can vary from up-coming prestige addresses to the inner city slums. Where slum clearance has occurred, huge apartment houses of thirty or more stories may have been built.

Further out, the suburbs begin. Duplex houses with yards front and back are common here, and the pace of life is less hurried than in the downtown areas. As the suburbs give way to the country, a few isolated farms are hidden among the trees.

The city's industrial activity tends to concentrate in the poorer outskirts, often close to the major freeways that link city to city, or run to the nearest docks from which goods can be shipped.

Freeways, railroads and subways span the city to carry workers to and from their work. Beneath the city streets, cables and pipes carry electricity, gas, water, telephone lines and sewage to service the city's buildings.

At work, at play or at home, city life depends on a continual and massive input of energy. But sources of energy are becoming increasingly scarce and expensive. An understanding of how energy is used in homes and cities, and of how it could be used more wisely, may be vital for the future.

The modern city is a hive of activity, day or night. Without a continuous supply of energy for its transportation, offices and homes the city would soon grind to a halt.

Power station 1
Power plants burn fossil fuels to supply electricity for the city's buildings.

Energy efficient house 6
The low-energy house is a suburban newcomer with a promising future.

Suburbs 7
Most suburban houses are owner-occupied and relatively spacious.

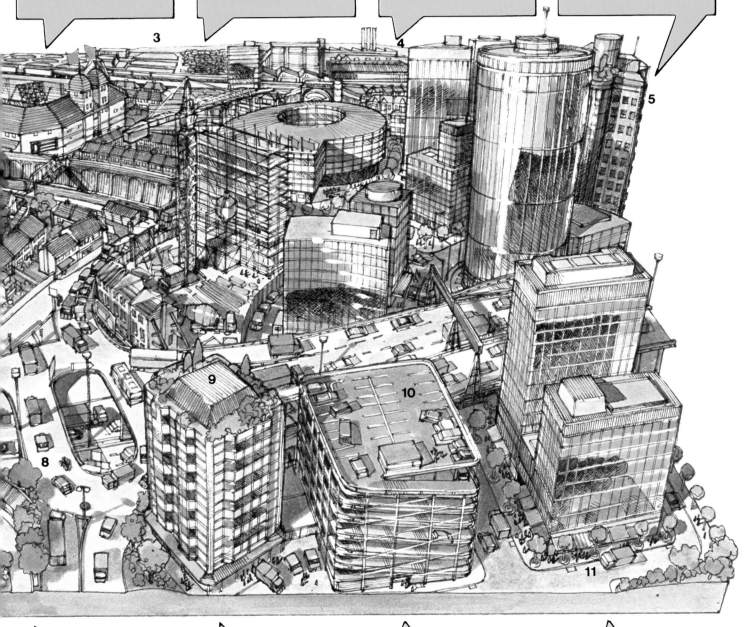

Sports stadium   2

Sports provide relaxation and entertainment for city dwellers.

Sewage   3

Treatment of sewage is vital in maintaining the public health.

Factories   4

Factories on the city's fringes provide people with goods and jobs.

Tower blocks   5

Prestige offices of large corporations give cities their dramatic skyline.

Transportation   8

Roads are the city's lifelines, along which people and goods flow.

Apartment block   9

Apartment houses accommodate a large proportion of the city's population.

Parking lot   10

Multistory parking uses a minimum of land to accommodate the city's autos.

Offices and shops   11

Downtown streets are lined with the city's stores, bars and restaurants.

Energy and cities

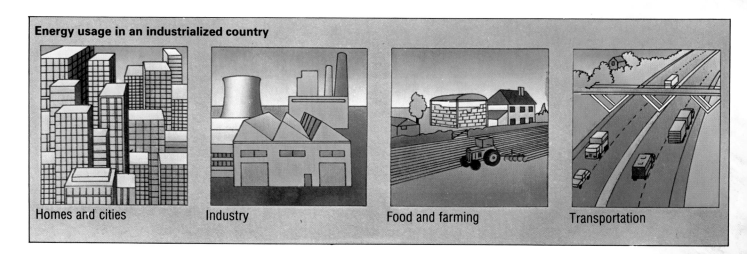

Energy usage in an industrialized country

Homes and cities | Industry | Food and farming | Transportation

Mankind has always been adept at obtaining energy from the environment around him. The discovery of fire brought heat to the first primitive caves and huts, and enabled sophisticated weapons and tools to be made. By harnessing animal power, or by working cooperatively, human beings developed the farming and building techniques which brought about the first permanent settlements.

Within the last 200 years, the tapping of the huge stores of energy available from the fossil fuels, firstly coal, and then gas and oil, has greatly increased the scope and scale of our activities. As industry grew, people began to flock to the cities to take the new jobs that were created. For the first time, city populations topped the one million figure, and went on climbing, using more and more energy as they did so.

Today's cities are totally dependent on large-scale energy supplies. Simply to construct a modern office block requires more energy than would have been used in a year in preindustrial cities.

Of course energy use doesn't begin or end with construction. Each of the separate materials that go into a building – steel for girders, cement and brickwork for the main fabric, and glass for windows – all have to be processed from the Earth's natural resources and transported to the site. The same is true for all the fittings that go into a building, from heating and air conditioning to desks and chairs. Even before the first occupants move in, a vast amount of energy has been used.

▷ In the USA, 35 percent of total energy use is consumed by offices and homes, 25 percent by transportation, 39 percent by industry and less than 2 percent by agriculture. Building a skyscraper requires advanced engineering skills. Often in poorer countries, only traditional local materials and manpower are used.

Once a modern building is in use it has an enormous appetite for energy. New York's World Trade Center, for example, uses as much energy in a year as does the entire city of Syracuse with almost 180,000 inhabitants. Without energy for heating, cooling, lighting and running elevators and other machines, buildings would be unable to function. Yet a lot of this energy is wasted, escaping into the atmosphere in the form of heat. Such heat losses account for a large amount of the energy used in homes and cities.

Buildings are not the only energy users in cities. The transportation, water supply, sewage disposal and communications systems that keep a city going all require energy for their construction and continued operation.

Cities in the developing world are just as dependent on energy, although overall they tend to use less than cities in industrialized countries. This is because a large proportion of their population have a very much lower standard of living. Consequently, although these countries contain about three-quarters of the world's population, they only account for an eighth of total world energy consumption. For many of these people, the only available sources of energy are the age-old ones of fuelwood, animal dung, and the power of human and animal muscles.

▷ This daytime photo of a group of Detroit offices was taken by a camera sensitive to heat rather than light. The lighter the area, the more it has been heated by the Sun. Air conditioning counters this effect. Similar shots taken at night show heat being lost.

Where people live

About 40 percent of the world's population lives in cities with 100,000 or more inhabitants. One in every hundred lives in the five most populous cities alone – Tokyo, Shanghai, Mexico City, Buenos Aires and Peking. The trend to city dwelling is increasing all the time, as people move from the country, and cities expand to swallow up surrounding towns and villages.

A quick check in the atlas will show that most of the world's major cities are located either on an important river, or on the coast. This is because they grew up as trading centers, importing and exporting goods to serve their surrounding areas. Inland cities, such as Teheran or Baghdad, are usually at the meeting point of two or more land or river trade routes. River valleys and coastlines also have the largest population density – much of India's teeming population lives in the fertile plain of the Ganges river.

In hot climates, rivers provide vital irrigation for food production. Indeed, a fertile soil, and plentiful irrigation are probably the two most important factors which help to determine where people choose to settle.

◁ Homes in suburbs, like these in British Columbia, Canada, tend to be more spacious than those in inner cities. On average, each of these houses will be occupied by a husband and wife and two children.

◁ New buildings get taller as land in New York's Manhattan gets scarcer and more expensive each year.

△ A village in the Kordofan district of Sudan, Africa. Each compound belongs to a large extended family.

Settlement in the USA

This map shows the USA seen from a satellite orbiting overhead. The lights shining from major towns and cities show that the most favored areas for human settlement are along the East and West coasts. Inland cities are relatively few and far between.

The type of environment people live in – city, town or village – determines to a great extent the type of dwelling they use to provide themselves with shelter. Apartment housing is common to cities worldwide. Multistory housing accommodates the maximum number of people in the limited and expensive land space available.

When people move into cities, their traditional patterns of living tend to be lost. The large multi-generation households that were once common in the USA and Europe have given way to homes in which only the parents and growing children live. With this trend has come a greater concern for individual privacy within the home.

In the smaller towns and villages of developing countries, the multi-generation dwelling is still the norm and personal privacy is almost nonexistent. Grandparents, parents, children and great-grandchildren, may all live under one roof, with only a rudimentary division into rooms inside the house. These homes may also be the family's workplace, or serve as shelter for valuable live-stock during the night.

11

Some simple rules

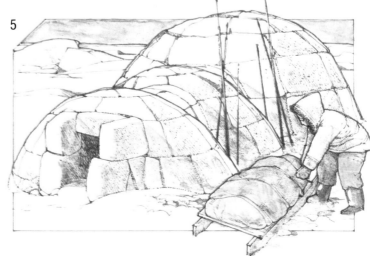

△ A goatskin tent of the Tuareg tribe (1) in the southern Sahara desert provides shade from the burning Sun. The tent sides do not reach the ground, so that cooling breezes can flow underneath. The Yagua Indians who live in the humid forests of Peru, achieve a similar aim by raising their open-sided bamboo houses on stilts (2). The whole tribe lives communally in just one of these buildings. For the nomads of the Mongolian steppes, the problem is to keep out the harsh winter cold. Their thick felt *yurts* are airtight (3).

△ Ingenious wind scoops on the roofs of these houses in western Pakistan direct cooling breezes into the rooms below.

▽ The Kirdi tribe live in the grasslands of northern Cameroon, Central Africa. Their huts have deep roof overhangs for shade and a thick layer of vegetation on top to absorb the Sun's heat (4). Eskimos live at the opposite extreme. The three chambers of their igloos (5) act as an air lock. Thick fur curtains between chambers help keep cold air out. Some of the world's highest temperatures are found in the Arizona deserts and houses here are built of thick adobe to create a cool, cellar-like interior (6).

3

Housing styles vary throughout the world and are often the product of generations of accumulated experience. Traditionally, people have used local materials and skills to create shelter in a sometimes hostile local climate. It is often possible to tell which part of the world a house is in, just by looking at its shape and the materials used to build it.

Dwellings from cold, windy areas, such as the Arctic wastes or the vast Asian steppes, are built using materials which slow down the rate at which heat is lost from the warm interior. The ice blocks Eskimos use to make igloos and the many layers of felt used in the Mongolian *yurt* do just this. The compact and often nearly hemispherical shape of these houses gives the maximum amount of space inside the building, and at the same time exposes a minimum area of outer wall through which heat can escape.

Hot dry climates pose different problems. In the Middle East and North Africa, houses usually have thick walls, made of sun-dried mud brick, or adobe. This soaks up the Sun's rays during the hottest part of the day, keeping the interior cool, and releases the stored heat during the night, when the temperature can drop dramatically.

Further protection is gained by shaping and positioning houses so that as little as possible of the outer surface is exposed to the fierce midday Sun. Windows facing the Sun are kept small, and external openings are placed to gain the cooling effect of local breezes.

While the structure of the dwelling, and the nature of the materials used to build it, afford shelter, they are not sufficient on their own to provide the comfort people require. Worldwide, people tend to prefer to live at temperatures within the range of 15–25°C (59–77°F), and clothing helps achieve this aim. No matter how well built the igloo is, an Eskimo still needs thick fur clothes to combat the intense cold of the Arctic. At the opposite extreme, in hot humid tropical climates, a minimum of clothing is most comfortable.

In industrialized societies these principles tend to be ignored. The clothing we wear is often simply a matter of fashion, and considerations of comfort and suitability to our climate take second place.

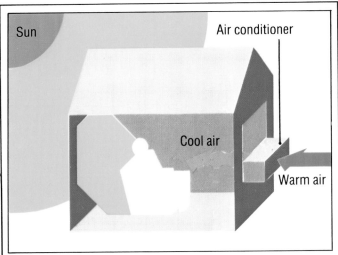

Sun

Air conditioner

Cool air

Warm air

Where the West goes wrong

The large window and thin walls of this house mean that the Sun will overheat it. The man inside has on too many clothes for comfort. He needs air conditioning to cool down incoming air and move it about his home.

6

Energy use at home

The typical home in industrialized countries is an intensive energy-user compared to the traditional housing of developing countries. Central heating and air conditioning are widely used to create a constant indoor temperature, winter and summer alike. At the touch of a button, we can transform a cold room into one warm enough to wear only a T-shirt and shorts. It is so easy, we seldom give it a second thought.

In fact, over half the energy used in a modern home goes into maintaining a comfortable temperature. Electricity, oil or gas may be used to power our heating systems, but the heat we pour in pours right out again through windows, doors, roofs and walls, making many homes, in effect, heat sieves. It is little wonder that many households in industrialized countries spend as much of their income on energy as they do on food.

It is not just living space that needs heating. Every morning we expect to be able to turn the faucet and get an immediate supply of hot water. Water heating accounts for 20 to 25 percent of home energy use, yet pipes and hot water storage tanks often have very inadequate insulation.

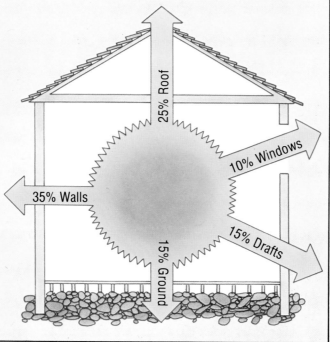

Heat loss in a Western house

An ordinary house in Europe or North America loses heat to the atmosphere during the winter. Outer walls and the roof are the main routes for escaping heat. Together they account for 60 percent of the heat lost. The remainder escapes through the windows, under doors and into the ground. When windows or doors are opened, heat floods out as cold air rushes in.

25% Roof

10% Windows

35% Walls

15% Drafts

15% Ground

△ In industrialized countries, even a simple meal such as breakfast uses up a good deal of energy. The stove, microwave oven, toaster and coffee machine are just some of the energy users within the home.

A Western kitchen

△ Fuel made from cow dung is as important to the households of this Indian village as electricity is to the typical American home. In areas where there are few trees it is often the only available fuel for cooking. It is burned in special clay stoves that retain heat for a long period. Although this practice serves an important need, the fertilizer value of the dung is wasted.

Energy use in the home does not stop there. Some rooms can have as many as four or five lightbulbs blazing away, and lights are often left on when they are not needed. We plug in and put to work a whole range of devices every day of the week. Stoves, freezers, vacuum cleaners, washing machines, dishwashers, garbage disposal units, coffee grinders – they all need energy and add to the household's fuel bills.

Few people in developing countries can afford this kind of extravagance. Electricity or oil and gas supplies are too expensive for most households to use. Daily chores, performed by machines in seconds in the richer countries, have to be done manually. For many people, the closest thing to a washing machine is the nearest river. Grains have to be ground by hand to make flour, and the only source of energy for cooking is an open fire. The difference in lifestyles between the world's rich and poor countries is reflected in the figures for energy consumption. The average American, for example, uses over 1,000 times the amount of energy used by the average Nepalese.

Saving energy

Insulating a cavity wall

Roof insulation

Saving energy

It is a relatively easy job to drill holes in outside walls and pump insulating foam, pellets, or fibers into the wall cavity. And anyone can lay glass fiber matting to insulate the roof. Hot water tanks and pipes should be given an insulating wrapping at the same time. The initial cost of these measures is soon recouped in resultingly lower heating bills.

Most of the houses in the industrialized world were built in a period when energy seemed cheap, plentiful and almost inexhaustible. Energy efficiency was not an important consideration in their design and construction. These houses still have a long useful lifetime left – in the UK, for example, 75 percent of today's houses will be in use in the year 2000. Measures taken now to cut down on the energy they consume will result in large savings over the years to come. Retrofitting older houses with better insulation and weather stripping, and improving heating systems could halve the amount of energy needed to heat a house.

Warm air rises, and a layer of thick, heat-retentive material in the loft can help prevent this heat escaping. Where walls are built with a cavity between two layers of brickwork, this cavity can be filled with insulating foam, pumped in from outside through small holes. Other insulating materials can be attached to both inner and outer walls, to further cut down on heat loss. In frame houses where the floor is made of timber, rather than heat-retentive concrete, it pays to insulate beneath, to prevent heat seeping into the ground.

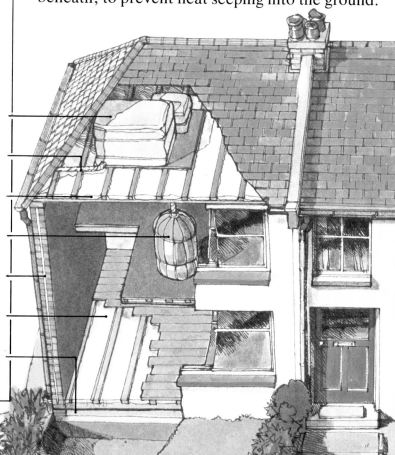

Insulated water tank

Insulated pipes

Roof insulation

Insulated hot water tank

Cavity wall insulation

Floor insulation

Damp proofing

Windows are another major escape route for heat. Losses here can be halved by double glazing, the air trapped between the two panes of glass being a poor conductor of heat. A valance above the window and a shelf below, and thick heavy drapes, stop the flow of cold drafts from the window into the room. Airtight weather-stripping can also prevent heat loss from windows and doors.

Improvements can be made in the heating system itself. Boilers, tanks and pipes need a thick insulation wrapping. Time switches ensure that heat is provided only when needed, and thermostatic valves on each radiator can be set to produce the exact desired temperature in each room. Solar panels on house roofs can meet half of the water-heating needs of the average family, even in relatively cloudy North European countries. These absorb the Sun's rays to heat the water passed through them. This water can then be stored in a well-insulated tank for later use.

The technology and materials for these improvements are already easily available in most countries, and all that is required is the will to apply them. Governments have an important role in encouraging these developments. A smaller fuel bill for each household means a smaller total energy bill for the country as a whole.

Solar panels for water heating

Double glazing

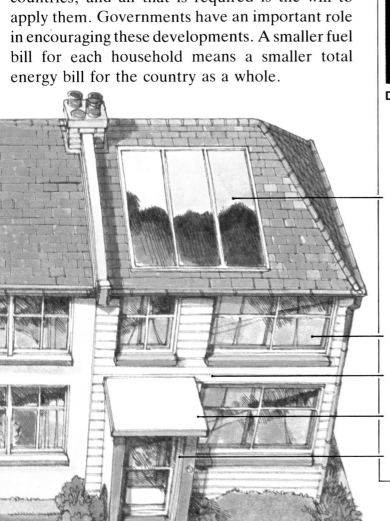

Solar panels

Double glazing

Extra external insulation

Stormdoor to stop drafts

Draft proofing

Making it warmer

More and more homes like these in London, have solar collectors on their roofs. These houses will gain all their hot water this way during the summer, but benefit little in the winter. In sunnier climates, year-round water heating is possible. Double glazing cuts heat loss from windows, especially if wood frames are used, as metal ones tend to conduct heat to the outside.

Low-energy use

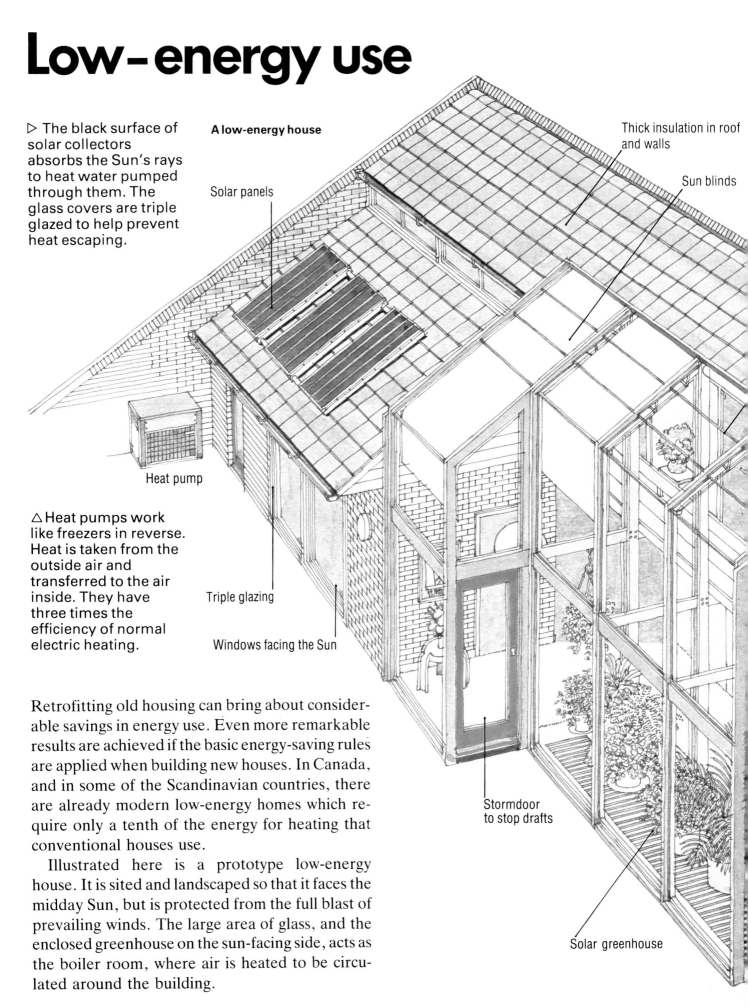

▷ The black surface of solar collectors absorbs the Sun's rays to heat water pumped through them. The glass covers are triple glazed to help prevent heat escaping.

A low-energy house

Solar panels

Thick insulation in roof and walls

Sun blinds

Heat pump

△ Heat pumps work like freezers in reverse. Heat is taken from the outside air and transferred to the air inside. They have three times the efficiency of normal electric heating.

Triple glazing

Windows facing the Sun

Stormdoor to stop drafts

Solar greenhouse

Retrofitting old housing can bring about considerable savings in energy use. Even more remarkable results are achieved if the basic energy-saving rules are applied when building new houses. In Canada, and in some of the Scandinavian countries, there are already modern low-energy homes which require only a tenth of the energy for heating that conventional houses use.

Illustrated here is a prototype low-energy house. It is sited and landscaped so that it faces the midday Sun, but is protected from the full blast of prevailing winds. The large area of glass, and the enclosed greenhouse on the sun-facing side, acts as the boiler room, where air is heated to be circulated around the building.

▽ The low-energy prototype house would gain 95 percent of its heat from the Sun. The air heated in the greenhouse rises up to the roof and is drawn into the loft. From here it is distributed to the rooms at the back of the house, and passes through a pebble bed under the floor, which stores the remaining heat to keep the house warm at night. The air circuit is completed when the now cold air is drawn back into the front of the house, to replace the rising warm air.

Heat losses from these large windows during the night are prevented by fitting blinds made of heat-reflective aluminized plastic. These blinds also help prevent overheating in summer, while the vents on the glass roof allow hot air to escape when the need arises.

In winter, further heating is gained by using a heat exchanger, by which incoming fresh air is heated by the warm stale air being expelled. In countries where the air temperature can fall as low as −20°C (−4°F), the soil is still a few degrees above freezing just 2 meters (6.5 ft) below the surface, and by passing fresh air through underground pipes a significant heating effect is gained.

A low-energy house such as this one would be too expensive to build to be within the reach of most people, but new buildings could incorporate at least one of the principles illustrated. The buildings being put up today will have a major role to play in saving energy in the future.

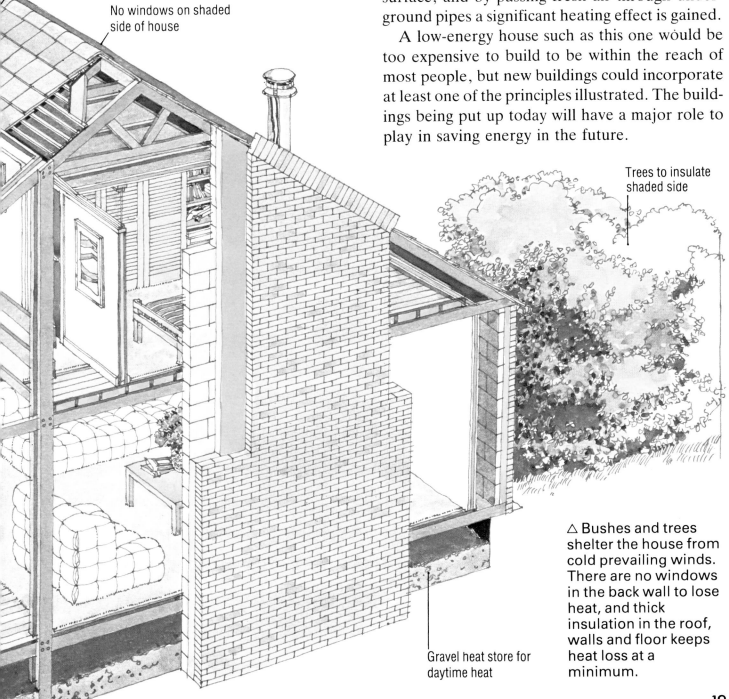

Air vents

No windows on shaded side of house

Trees to insulate shaded side

△ Bushes and trees shelter the house from cold prevailing winds. There are no windows in the back wall to lose heat, and thick insulation in the roof, walls and floor keeps heat loss at a minimum.

Gravel heat store for daytime heat

19

The hidden cost

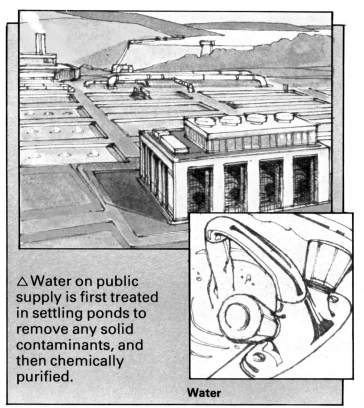

△ Water on public supply is first treated in settling ponds to remove any solid contaminants, and then chemically purified.

Water

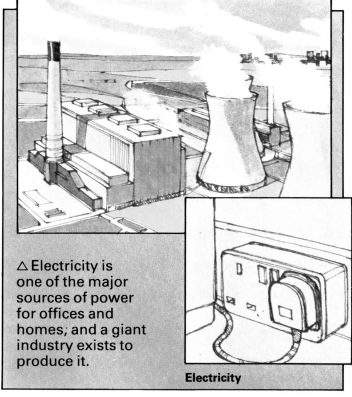

△ Electricity is one of the major sources of power for offices and homes; and a giant industry exists to produce it.

Electricity

No individual building, large or small, old or new, can be totally self-sufficient for its needs. Each is connected to a number of vast and complex support systems, which help make the everyday life of industrial societies possible. If any one of these were to break down, life in our homes and cities would be seriously disrupted.

Perhaps the most important of these are the energy supply systems, since all the other systems, water supply, sewage disposal and communications, depend on energy to function.

The power points in homes and offices allow us to plug into an electric supply grid that is often national in scale. Electricity to feed the grid is produced by burning fossil fuels in massive power plants. Even further down the line, these fossil fuels have to be extracted from the Earth, processed and transported often thousands of miles.

Similar systems exist to process and distribute natural gas used for heating and cooking. Buildings such as hospitals and schools, which have oil-fueled heating systems, depend on regular deliveries of oil to fuel their boilers.

△ Modern homes and cities depend on systems that serve thousands, and often millions, of people. Without energy, often in the form of electricity, these systems themselves would soon cease to function.

▷ An aerial view of Basildon, near London, shows the wider, less obvious services to which a home is connected. We need industries to supply our goods, road and rail services to get from one place to another, and schools, hospitals and libraries. Many of these services are paid for indirectly, by taxation.

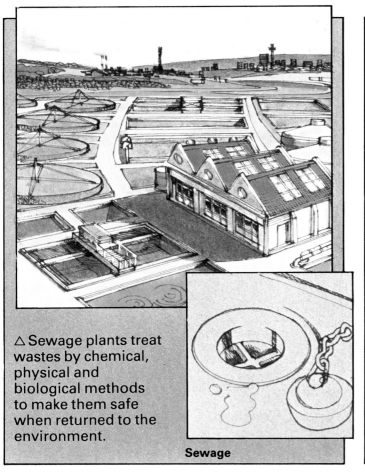

△ Sewage plants treat wastes by chemical, physical and biological methods to make them safe when returned to the environment.

Sewage

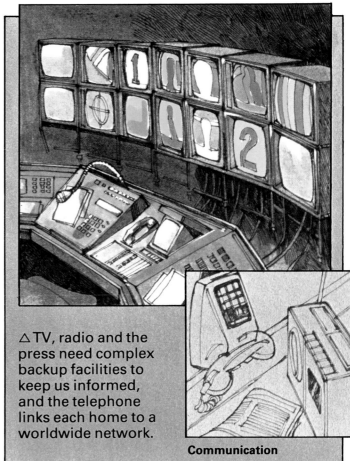

△ TV, radio and the press need complex backup facilities to keep us informed, and the telephone links each home to a worldwide network.

Communication

Homes also need a water supply, for washing, toilets and cooking, and for health reasons that water has to be clean. In hot, affluent countries such as the USA, or Australia, water usage can reach an average of 500 liters (132 gallons) per person per day. A huge network of reservoirs, purification and pumping plants, and hundreds of miles of pipe, is needed to satisfy this demand.

When the bathtub plug is pulled, the used water flows into the sewage waste pipe and out of the house to the nearest sewage treatment plant. Sewage treatment and disposal is vital – in poorer countries untreated sewage spreads diseases such as typhoid, cholera and diphtheria. These diseases no longer exist in countries where each home is connected to a sewage disposal system.

The communication systems, such as the mail and telephones, may seem less important than others, but modern society is too complex to function without them. Similarly, no household can exist in isolation from other aspects of society. Hospitals, schools, public transportation and libraries are essential to our everyday life.

City systems

To move from the home to the city is to move into a larger scale of energy use. Each day, hundreds of thousands of people commute into the world's major cities. One square mile of Manhattan, for example, experiences an influx of over three-quarters of a million workers every morning. Moving such a vast number of people requires a highly organized and energy-consuming transportation network of autos, buses, trains and subways.

In the past 40 years, the number of autos in use has increased twenty fold. Traffic jams on the major highways in and out of cities during the rush hours are familiar to every driver, and the congestion gets worse every year.

Many cities are encouraging commuters to switch to public transportation. A new rapid transit system, BART (Bay Area Rapid Transit), was completed in 1974 to serve San Francisco, and other cities have invested in modernization or a low fares policy for their public transportation. These systems are usually faster and cheaper to use than automobiles, and overall they consume only half as much per passenger mile.

◁ The lights burning in these Tokyo skyscrapers outside of office hours waste energy. The slab-like shape of these buildings means heat is lost to prevailing winds, as a large area of outer wall is exposed.

▷ Firemen tackling a blaze in Germany. Cities need emergency services to cope with disasters such as fires, railway accidents, or even earthquakes. These services are usually paid for out of local taxes. If fires were not controlled, they could devastate city centers, as has often happened in the past.

◁ San Francisco's Bay Area Rapid Transit system (BART) is one of the world's most advanced commuter railroads. It has 114km (71 miles) of track, with a 6km (3.7 miles) section running underwater across San Francisco Bay. The whole system is computerized, and controlled by just two staff in the central control room.

Anyone who has flown over a city at night will have been struck by the way that it stands out as an island of light in the surrounding darkness. The city streets blaze with neon signs, street lights and illuminated shop windows. But the biggest urban users of lighting are the skyscrapers, where lights are often left on day and night.

Not all the city systems are as visible as transportation and lighting. No city can function without the emergency services – fire, police and ambulance – that swing into action as soon as anything goes wrong. If streets are not cleaned, parks maintained and garbage collected, city-dwellers are quick to complain. None of these activities could take place without a well-developed system of municipal, federal and national administration. This is probably the most important system of all, as it determines how well, or how badly, a city is able to adapt to changing conditions – such as the rising price of energy.

△ The UK's Central Electricity Generating Board sets an example with its office complex in Bristol. It was designed to use a minimum of energy, and the need for electric lighting is reduced by allowing the maximum amount of daylight in through skylights and windows.

Providing power

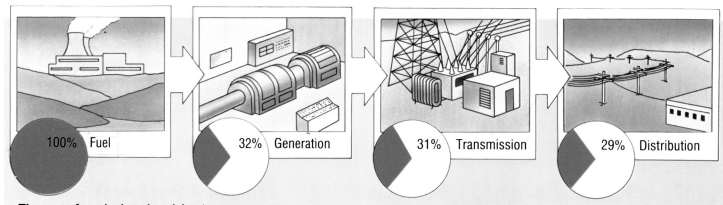

The cost of producing electricity (Percentages above refer to energy remaining after a particular process)

100% Fuel	32% Generation	31% Transmission	29% Distribution

Electric power plants are large energy consumers themselves. In the first place, energy is needed to obtain the fuels they use from the earth and to transport them to the plant. Only 32 percent of the energy value of the fuel is converted to electricity in the generating process. To enable the electricity to be transmitted long distances, the voltage is raised by a factor of ten or fifteen times in a transformer. This has to be reduced again to 220 volts before it can be used to run home appliances safely.

Electricity is one of the most easily available forms of energy used in homes and cities. In industrialized countries electricity production accounts for around 30 percent of the total fossil fuel usage. However, current methods of converting the energy value of fossil fuels into electrical energy are highly inefficient, and this makes electricity one of the most expensive forms of energy.

In conventional power plants, fossil fuels are burned to convert water to pressurized steam. This steam turns the generators that produce electricity. In the course of this conversion over two-thirds of the fossil fuel's energy value is lost in the form of waste heat vented out to the atmosphere. Even the most advanced nuclear power plant is little more than an atomic kettle, in which energy from uranium atoms is used to make steam. The same conversion losses also apply to these plants.

This waste heat can be used to heat water to be piped off to local homes and offices. This is being done on an increasing scale in Scandinavia and North America. Such cogeneration plants, as they are called, result in over 70 percent of the energy value of the fossil fuel being recovered – over twice as much as in conventional plants.

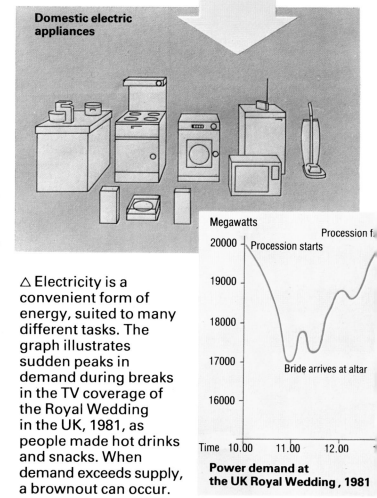

Domestic electric appliances

△ Electricity is a convenient form of energy, suited to many different tasks. The graph illustrates sudden peaks in demand during breaks in the TV coverage of the Royal Wedding in the UK, 1981, as people made hot drinks and snacks. When demand exceeds supply, a brownout can occur.

Power demand at the UK Royal Wedding, 1981

Combined heat and power

The streets of Helsinki, Finland, like those in some Danish and Swedish cities, are being excavated as part of a cogeneration and district heating project. The pipes laid will carry hot water from local power plants to schools, homes and offices.

Totem

The Totem unit, based on a Fiat 127 auto engine, can generate up to 15kw of electricity. In addition, waste heat is reclaimed by passing water through a heat-recovery jacket around the engine. This water can then be piped to supply local buildings.

A single cogeneration plant in Odensee, Denmark (population 130,000), supplies heat to 90 percent of the households there, through over 1,000 km (621 miles) of pipe. Its electricity output feeds straight into the national grid.

In district heating, low-grade fuels, and even domestic wastes, are burned to provide local buildings with heated water, further reducing the need to use electricity for this purpose.

The same principles can be applied on a smaller scale. The Italian automobile manufacturers, Fiat, have produced an electricity and hot water supply unit – the Totem – based on a modified auto engine. The Totem can use a variety of fuels, from gasoline to the biogas produced from fermented agricultural waste, and one unit can supply heat and electricity to at least five households.

Fuel cell power plants which can directly convert the energy of chemical fuels into electricity, are currently being developed. These work in a similar way to ordinary batteries, but on a much larger scale. The most modern fuel cells are three times as efficient as present electricity generation methods, losing less than 20 percent of the energy value of the fuel used.

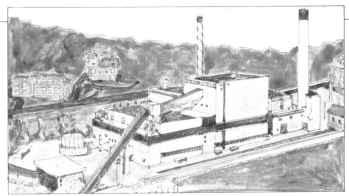

Mixed fuels

△The Naistenlahti combined heat and power plant in the city of Tampere, Finland, is fueled by both peat and oil.

Fuel cell

▽ Hydrogen-rich gas is extracted from fuel and is then chemically reacted with air in the power section of a fuel cell to produce electricity.

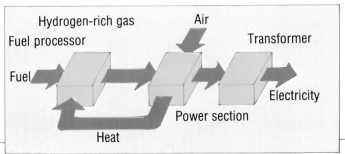

A national solution

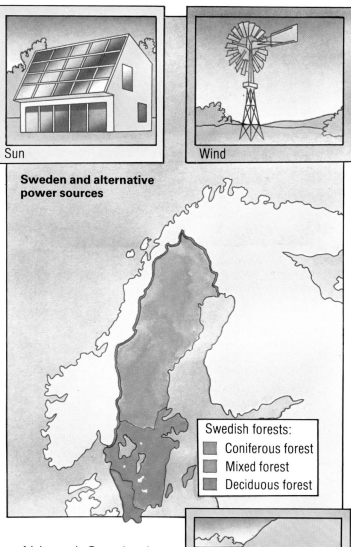

Sun

Wind

Sweden and alternative power sources

Swedish forests:
- Coniferous forest
- Mixed forest
- Deciduous forest

Water

Sweden was one of the first industrialized countries to develop an energy policy for its homes and cities in response to the problems posed by the ending of the era of cheap energy. In 1973, when the price of oil began to soar, Sweden got over 80 percent of its energy from imported oil. Sweden hopes to halve that figure by the end of the 1980s, while maintaining its high living standards.

Sweden has ruled out the nuclear option – a public referendum decided that no new nuclear plants were to be built, and existing ones were gradually to be phased out. Hydroelectric power from Sweden's many lakes and rivers already provides 65 percent of the country's electricity. Wind and wave power from around Sweden's coast will make an increasing contribution to the electricity supply in the coming years.

Over a third of the country's oil goes to heat buildings, and an extensive public campaign has been mounted to improve their insulation and energy-efficiency. New technologies, based on alternative fuel sources, are being explored and developed. The geothermal energy of the hot rocks beneath the Earth's surface is already being used to provide a hot water supply in the southern part of the country.

△ Although Sweden is a northern country stretching beyond the Arctic circle, it gets enough sunlight to make solar power a major contributor to its energy policy. Large areas of the country are forested and can provide Sweden with biological fuels to substitute for oil. Sweden's mountain rivers and lakes are already providing hydroelectricity at almost full capacity, while wind and wave power have great development potential.

▷ A Swedish professor, Dr. Wolgast, has built an experimental low-energy house which he lives in himself. With thick insulation and the recovery of waste heat, the house needs no heating when the outside temperature is above −5°C (23°F).

Wolgast's house

Much of Sweden is covered in forests, and the Swedes are planting fast-growing tree crops, such as the poplar, to use as an alternative fuel to oil. The poplar can be harvested after just three years, and the whole tree, leaves and all, is shredded, dried and then densely compacted into solid fuel pellets. These pellets are much more convenient to handle and transport than traditional wood logs, and contain a much higher energy value for a given weight. The Swedes are currently researching methods to convert the energy value of poplar into a liquid fuel which could be put to a much wider range of uses.

Solar heating is being applied not just in individual houses, but on a community scale. Many homes, offices and schools are now partially or totally heated by the sun's rays, and a project to convert the picturesque university town of Lund entirely to solar heating has recently begun. In many of these projects, large reservoirs of hot water are used to carry the extra heat collected in summer over for winter use.

Many modern Swedish buildings use less than half as much energy on heating as comparable buildings in milder climates, such as the UK. Sweden's bold and imaginative energy policies will cut this requirement further, and Sweden could soon become the model that other industrialized countries will choose to follow.

△ Solar collectors at Studsvik rotate to follow the Sun. They are attached to the floating lid of an insulated hot water reservoir, which will be used by nearby offices and other buildings for winter heating.

▽ Six houses at Bramhult are heated by solar energy. Any surplus heat is stored in a communal reservoir, monitored by computer for maximum efficiency, to be used in the cold winter months.

Bramhult solar community

An alternative answer

The high cost of energy is a major problem for many poorer developing countries, and contributes to their lower standard of living. China has, for the past three decades, pursued a path of low-cost alternative energy usage that many other developing countries, facing similar problems, could easily adopt to their advantage.

China's energy policy is applied at all levels, from small village or commune, to the big cities. To make the best possible use of their resources, the Chinese have integrated their industrial and agricultural developments. Agricultural and domestic wastes are seen as a major source of fuel, so much so that the Chinese don't refer to them as waste at all, but as precious treasures.

Early every morning, domestic sewage is collected from Chinese towns and transported to the surrounding food-growing areas. It is then composted to kill off any disease-carrying organisms, before it is returned to the soil as a valuable fertilizer. In this way, three important demands are fulfilled. Sewage is hygienically disposed, the demand for water supply is reduced by eliminating flush toilets, and agricultural production is increased, to help feed China's huge population.

△Low-cost biogas digesters being built and installed in Sichuan province. Each digester will supply a household with biogas for cooking and heating purposes.

◁ A typical hand-pulled cart carrying night sewage from the city of Chengdu to a nearby digester for biogas production. Manpower is commonly used to pull vehicles throughout China.

▷ A fish farm at the Jinma commune, in Sichuan province. A dependable water flow is produced by local irrigation and flood control schemes. The building itself will use biogas from waste.

▽ The plastic bag on top of this two-wheel tractor is the vehicle's gas tank – it contains biogas. A standard engine needs a slight modification to run successfully on biogas. Other farm machinery can be fueled in this way.

In the rural areas of China, the energy value of animal dung is fully realized. It is composted in sealed containers, known as digesters, to produce biogas, which is similar to the natural gas extracted from the Earth. The biogas is pumped to homes and factories, to be used for cooking and heating, but it can also be used to power vehicles and farm equipment. The liquid slurry left over after gas production is rich in fertilizers, and is again returned to the earth. There are currently over four million biogas plants in use in China, most of which serve single households, although some are on a larger scale, providing energy for whole communities.

In China, ditches, canals and dams, primarily for irrigation purposes, are also used as energy sources. On the largest scale, giant dams can be used to produce hydroelectricity for China's major cities. Local rivers in some rural areas can generate enough electricity to run a small manufacturing plant. China's network of waterways also forms a low-energy transportation system using barges to carry necessary goods, and many of the lakes behind dams serve as fish farms, providing another valuable source of protein.

Pressures and problems

In cities all over the world, literally millions of people are packed into a relatively small area of land – often as many as 10,000 per square kilometer (0.39 sq. miles). As populations increase and the cost of energy rises, cities worldwide are experiencing ever-increasing pressure.

Unemployment and poverty affect large sections of city populations. Every major city has its ghetto areas, where old buildings have been allowed to decay, inhabited by those who are too poor to make improvements or move elsewhere.

Recently, many poorer households have found it difficult to meet payments for their gas and electric supplies. This fuel poverty, as it is called, is a growing problem in many European and North American cities, with increasing numbers of households being disconnected from their fuel supply. For the very young and the old, inadequate heating in winter can be fatal.

The pressures of poverty and inner city decay lead directly to social problems. Cities are becoming increasingly tense places in which to live. Crime rates are rising, and cities in both the USA and the UK have seen the tensions in poorer urban areas erupt into riots in the past decade.

△ Rush hour on the crowded Tokyo subway, with commuters being packed tighter than sardines in a can. The stress on the transportation network itself, as well as that on passengers is considerable.

◁ Barges take garbage collected in central London along the River Thames to the estuary marshes, for use in land reclamation. Similar schemes are operated in San Francisco. Cities worldwide need all the space they can get.

The direct result of the decline in the quality of urban life is that those who can afford to move out of the city do so. The population of London fell by around two million in the past 40 years – about a fifth of the city's total population. The same trend is happening in all the major cities of the industrialized world.

Cities in developing countries face even harsher pressures of poverty, but added to this is the problem of an explosive population growth. Mexico City, for example, in contrast to London, has increased its population by about two and a half million in just the last ten years.

This growth is not simply due to the high birth rates in developing countries. The myth that city streets are paved with gold is common to all countries. Worsening rural poverty leads to a mass exodus away from the countryside into the city, as people seek greater opportunities. Many of this new influx settle in shanty towns with no electricity or water supply and no sewage disposal system. Others literally live on the streets – Bombay alone has a street population of over 100,000 people. For these people, the urban poverty of industrialized countries seems almost luxurious.

△ The slums of Harlem in New York typify the worst of inner city decay, although similar scenes can be found in other North American and European cities. It is not just buildings that become derelict, but whole communities, as people become trapped in a vicious circle of unemployment, poverty and crime.

▷ Bombay is the Hollywood of India, the center of a wealthy film industry, but life for most of its inhabitants is far from glamorous. Other cities in the Third World show similar contrasts as populations increase.

Which way?

Until recently, there was a tendency to think that many of our problems could be solved by building on a larger scale. Industrialized society's increasing demand for energy, it was thought, could be met by massive power plants, consuming even greater amounts of fossil fuels, or by developing nuclear energy as an alternative. It is only since the oil crisis of 1973 that people have started to think in terms of energy conservation, and of reducing our energy requirements.

Rising energy costs are not the only factor that has led to this readjustment. Giant nuclear plants are vastly expensive, and most countries simply cannot afford to build the number required to meet the electricity demands of the future. Technical failures, such as the one which caused the New York blackout in 1979, stranding people in elevators and subways, have made us question the wisdom of large-scale energy dependence, and the accident at the nuclear plant at Three Mile Island renewed fears about the safety of nuclear power.

△ A solar-heated pool is a luxury few people can afford, but it is one which does not make any demand on our diminishing fossil fuel resources.

▽ Self-sufficient houses are hard work. People have to get up early to tend to the livestock, as well as maintain their windmills and solar collectors.

Self-sufficient housing

Energy supply is not the only area of increasing scale in modern life. The Sears Roebuck Tower in Chicago, currently the world's tallest building, soars to an astonishing 443 meters (1,454 ft). Architects have designed apartment housing with anything from twenty to seventy stories, to replace slum tenements. These giant apartments may save limited city space, but they are often unpleasant places in which to live.

The reaction against this trend to make everything larger was summed up in the 1970s by the phrase "small is beautiful." As an extreme alternative, experimental self-sufficient households were suggested, and given much publicity. These households would provide their own energy and water supplies, and deal with their own waste disposal. A fishpond, greenhouse and livestock would provide self-sufficiency in food.

Of course it is just as undesirable for every household to become self-sufficient as it is for cities to be gigantic. Many things are most efficiently done on a community scale – a large heat storage tank serving a whole block uses less energy and fewer materials for its construction than thirty or forty smaller ones. The better alternative is to choose the right scale, to meet the differing needs of different communities.

△ The monotonous face of an apartment house in Glasgow, Scotland. Such multi-story housing has been a social failure, and some projects, built only ten years ago, have been demolished.

◁ A society organized on a large scale has to be a high energy-consumer. New electric plants, like this nuclear one at Oldbury, UK, have been built to meet our increased energy demands. Such projects require a massive initial capital investment.

Breaking point

Science fiction writers often depict future cities as daunting, highly technological landscapes, where little is on a recognizably human scale. In these energy-intensive mega-cities, a vast web of super freeways crisscrosses between the mile-high skyscrapers. Inside these buildings the environment is kept at a comfortable norm all year round. Outside, perhaps, a thick layer of smog hangs in the air, but people in this vision of the future seldom venture outdoors any more.

Of course, such a society would require huge amounts of energy to keep it going, and this could theoretically be supplied by fusion power. Fusion is the way in which the Sun generates its energy – a fusion power plant would be like a mini-sun here on Earth. The scientists now researching fusion are faced with the problem of containing the enormous amounts of energy it produces, but this could be solved within the next century.

Until that happens, the energy-intensive future will probably have to depend on coal and nuclear power. There is enough coal left in the Earth to supply us for a hundred years or so, even if we did consume it in vastly increased amounts.

A city of the future?

▷ Twenty more years of intensive energy use could leave our cities looking like this. Giant freeways are needed to cope with increased traffic, and living space has become more cramped than ever. Pollution, in the form of acid rain and smog, from autos and trucks is made worse by the burning of coal, the major energy source along with nuclear power. Over ten times as many nuclear plants as we have now will be needed to meet the increased energy demands of industry and cities.

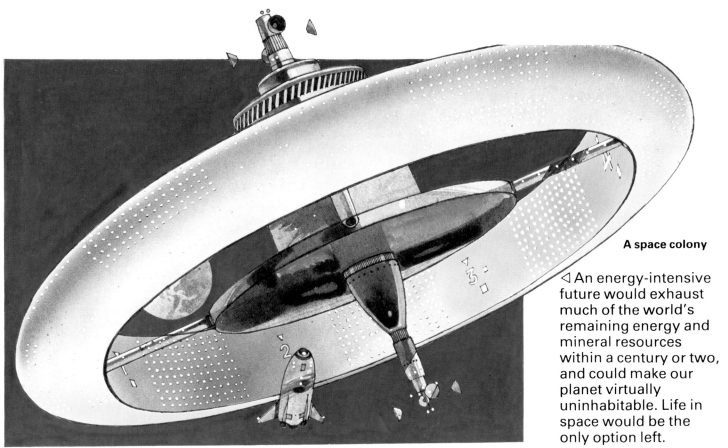

A space colony

◁ An energy-intensive future would exhaust much of the world's remaining energy and mineral resources within a century or two, and could make our planet virtually uninhabitable. Life in space would be the only option left.

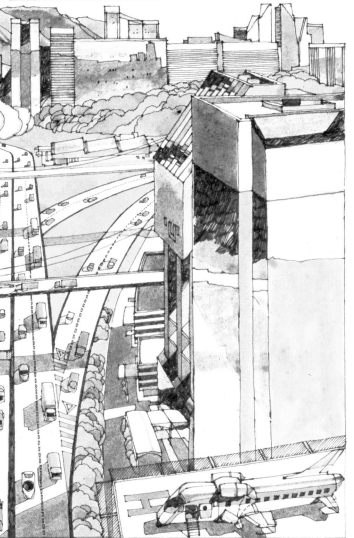

If such an energy-intensive future became reality, it could have some very unpleasant consequences. First, by burning more coal, greater amounts of carbon dioxide would be released into the atmosphere. This acts as a kind of heat blanket, reflecting the heat produced by massive energy use back to Earth. Normally, trees and other plants absorb carbon dioxide and convert it to oxygen, but in the high-energy future, most of the forests will have been cleared for building space and food production. The net effect would be a worldwide rise in average temperature. Some climatologists believe that a rise of as little as 2°C (4°F) would turn the grain-growing areas of the USA's Midwest into an arid desert, causing world famine. A further increase of the same order would melt the polar icecaps, causing sea levels to rise and flood most of the world's populated areas.

The final option for such a society would be space. In the distant future a select few might be able to abandon spaceship Earth in spaceships of their own making. We have already seen men on the moon while for much of the world's population it is hard to find fuel to cook the daily meal. An answer to our energy problems which ignores most of mankind is no answer at all.

A possible future

It is very unlikely that industrialized societies will go all out for an energy-intensive, highly technological future. Economic, social and environmental pressures would soon prove intolerable. Equally, those who enjoy the comforts and benefits of life in modern homes and cities would be unwilling to go back to a totally self-sufficient, rural way of life.

The probable pattern for the future is a middle course between these extremes. The burning of fossil fuels to generate electricity will be phased out, to be replaced by fuel cells, windmills, hydroelectricity, and power plants that harness the energy of waves. Electricity will be reserved for uses to which it is most suited – such as for telecommunications – rather than for heating.

Local communities will use heat generated from burning their own waste, and from solar collectors, to cut down their energy demands. In areas where the soil is too poor for food cultivation, trees will be grown to be processed into liquid or solid fuel. Seaweed from coastal areas will be exploited for its fuel value in the same way. More and more, these biological fuels will be used instead of oil and gas in our homes and cities.

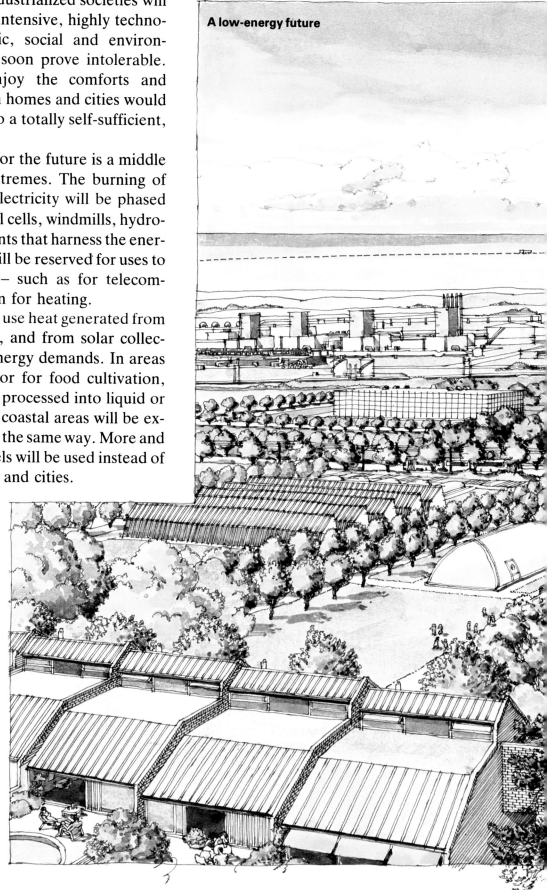

A low-energy future

▷ In twenty years' time, the technology necessary to tap the renewable sources of energy – wind, wave and solar power – will have been developed and applied. Individual buildings and small communities will be heated partly by solar collectors and partly by local cogeneration and district heating plants fueled by a combination of coal, biological fuels, and domestic wastes. The standard of living will be high, and the outside environment free from harmful pollution.

Other areas of life will be affected by the energy rethink. Automobiles and trucks will be designed to use less fuel for a given mileage, but their use will be less common than it is now. An integrated public transportation system will serve cities, thus cutting fossil fuel usage, and making our cities less hectic places in which to live.

Goods that do not need rapid transport will travel, where possible, by river and sea, using sophisticated wind-powered ships and barges. As production becomes more localized, fewer goods will actually need long-distance transportation. As new computer technology brings about improvements in telecommunications, much of the need for business travel will be removed.

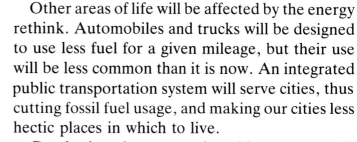

These developments would not require any fall in peoples' living standards. A well-insulated house is more comfortable than one with drafts whistling in at every window and door. The savings made, in both money and resources, by *not* squandering energy could be used to benefit us in other ways – by clearing up our city environments, for example, or by helping developing countries to solve their many problems.

Such a future will not happen automatically – it requires governments to have the determination to encourage these changes. It also requires individuals to reexamine their attitudes to energy use in their own homes and places of work. It is everybody's responsibility, and in everybody's interest, to see to it that we can look forward to the future with reasonable optimism.

Index

Acknowledgements
The publishers wish to thank the following people who have helped in the preparation of this book: Architectural Association; BP Chemicals; Central Electricity Generating Board; Department of Energy; Department of the Environment; District Heating Association; Earthscan; Electricity Council; Energy Research Group; General Electric Company; National Swedish Board for Energy Source Development; Royal Institute of British Architects; Royston Summers; Swedish Council for Building Research; Swedish Embassy; William Orchard & Partners.

Photographic Credits
In cases where more than one photograph appears on a page, credits are listed from top to bottom and from left to right. Page 9, Daedalus Enterprises; page 10, Canadian High Commission, Zefa; page 11, Earthscan/Mark Edwards; page 15, Earthscan; page 16, National Cavity Insulation Association, Cape Insulation; page 17, SLC Energy Group, Glass & Glazing Federation; page 20, Aerofilms Ltd.; page 22, Zefa, BART; page 23, Zefa, CEGB; page 25, District Heating Association, Fiat Energy; page 27, Studsvik; page 28, Peter Fraenkel; page 29, Peter Fraenkel; page 30, Zefa, DOE; page 31, Architectural Association, Alan Hutchison; page 32, Architectural Association; page 33, Network, CEGB; Endpapers, George Wimpey & Co.